CONTENTS

© T.F.H. Publications, Inc.

Distributed in the UNITED STATES by T.F.H. Publications, Inc., 1 TFH Plaza, Neptune
City, NJ 07753; on the Internet at www.tfh.com; in CANADA by Rolf C. Hagen Inc.,
3225 Sartelon St., Montreal, Quebec H4R 1E8; Pet Trade by H & L Pet Supplies Inc.,
27 Kingston Crescent, Kitchener, Ontario N2B 2T6; in ENGLAND by T.F.H. Publica-
tions, PO Box 74, Havant PO9 5TT; in AUSTRALIA AND THE SOUTH PACIFIC by
T.F.H. (Australia), Pty. Ltd., Box 149, Brookvale 2100 N.S.W., Australia; in NEW
ZEALAND by Brooklands Aquarium Ltd., 5 McGiven Drive, New Plymouth, RD1 New
Zealand; in SOUTH AFRICA by Rolf C. Hagen S.A. (PTY.) LTD., P.O. Box 201199,
Durban North 4016, South Africa; in JAPAN by T.F.H. Publications. Published by
T.F.H. Publications, Inc.

MANUFACTURED IN THE
UNITED STATES OF AMERICA
BY T.F.H. PUBLICATIONS, INC.

D1339751

INTRODUCTION

Cichlids are among the most popular aquarium fish. Many of them are large, tough, mean, destructive, and predatory, and while some aquarists like them because of these traits, most of difficulty to the breeder, from the simplest easy-for-beginners like the convict cichlid, *Herichthys nigrofasciatus*, to the super challenges like the discus and the uaru, *Symphosodon*

Many of the dwarf cichlids are much easier to keep and breed than larger cichlid species such as this *Uaru amphiacanthoides* . Photo by Edward Taylor.

enjoy cichlids despite their large, tough, mean, destructive, and predatory natures. They like them because they are beautiful, intelligent, interesting, and rewarding aquarium pets and because they offer a full gamut *discus* and *Uaru amphiacanthoides*. The species known collectively as dwarf cichlids provide many of these virtues without the rougher edges.

Some people think of "big cichlid" aquarists as bullies who

live vicariously through their pets, which require a 150 gallon tank for each pair. They are seen as filling their houses with giant aquaria into which they pour endless quantities of feeder goldfish, then jump back from the splashing and watch with glee.

The equivalent image for "dwarf cichlid" aquarists is namby pambies who putter around, siphoning live baby brine shrimp into racks of five gallon tanks that often seem to house only some peat moss and a broken flowerpot.

Well, as wrong as these stereotypes about cichlid enthusiasts are, all the stereotypes about the cichlids themselves are blasted apart by one or more species. In this book we will examine the notion of "dwarf cichlids," its limits and its usefulness, then talk about the most commonly encountered species and provide a guideline for their care and breeding.

Whether you are relatively new to the hobby or an experienced aquarist who wants to learn more about these little gems of nature, the information here will enable you to start enjoying dwarf cichlids in your own home.

Lamprologus caudopunctatus is a cave-spawning jewel from Africa. If you can find them, buy them. Photo by Bob Allen.

WHAT ARE DWARF CICHLIDS?

Behavior figures as prominently as size in most people's minds when defining the dwarf cichlids. The whole thing is complicated by the fact that not everyone agrees on which are which, or by what criteria they should decide. Unfortunately, there is no family Nanocichlidae to which all dwarf cichlids are assigned. Nor is there any set of attributes, including adult size, which will correctly lump dwarf cichlids into one group apart from all other cichlids. Most people think of dwarfs as colorful, peaceful, diminutive, sexually dimorphic species that spawn in caves. But will that work?

Well, a mature convict can be shorter than a krib, but only because the smallest convict size overlaps with the largest krib size. Also, the convict is a stocky fish, a scaled down big cichlid, while the krib is a slender, petite fish. In addition, the krib is a whole lot more "dwarf" in nature than the assertive convict, and while both will breed in a flowerpot, for the convicts it is a stronghold from which search and destroy missions are waged, and the area around it is often pockmarked with craters dug by the busy parents. For the kribs, the flowerpot is a secret refuge in and around which the parents tend their fry. In both species the female is more colorful, getting a brightly colored belly blotch during breeding, and her mate normally outsizes her considerably.

A male convict cichlid, *H. nigrofasciatus*. This species is easily spawned and zealously guards its eggs and fry. A good fish for beginners. This is a South American fish. Photo by Hans Joachim Richter.

The ram, *Microgeophagus ramirezi,* guarding his eggs which were laid and fertilized on a flat rock. This fish comes from South America. Photo by Has Joachim Richter.

Or consider the diminutive African rift cichlid *Lamprologus caudopunctatus*; it is a cave-spawning, slender little jewel, but it can hold its own against fish many times its size, and many people would call it a small, aggressive cichlid rather than a dwarf cichlid.

In temperament and behavior, the nervous and peaceful angelfish and discus are, despite their large size, much more similar to the *Apistogramma* than to Jack Dempseys, Oscars, and Red Terrors.

Some very large cichlids breed in (very large) caves, but some tiny ones, like the ram, *Microgeophagus ramirezi,* spawn in the open, and medium ones, like the convict, spawn either way.

The Asian chromide is a smallish cichlid, brightly colored and unusually peaceful, but not really dimorphic. It is a ready breeder and a good parent, and while its sunfish shape makes it look like a small relative of the severum, its deportment is much like that of the unanimously accepted true dwarf cichlids, and many people include it with them.

You can see that the notion of "dwarf cichlid" is a vague one, hard to define, and while most aquarists can agree on a group of species as all being dwarfs, and on a group they all consider not to be dwarfs, there is a large grey area of disagreement.

Actually, it would be more accurate to say that there are many cichlid species for which the dichotomy of "dwarf" versus "big" cichlid is not valid. The rift lake species, for example, are in many ways best considered together, both large and small.

Part of the reason for this confusion is historical. Years ago, the limited availability of cichlid species gave the impression that there was a taxonomic grouping of "dwarf cichlids," which shared certain traits not belonging to regular cichlids. Cichlids available to hobbyists were divisible into two basic groups: big ones and dwarfs. The large ones were represented by the Central American *"Cichlasomas"* or the South American Oscars, discus, angelfish, and acaras. The dwarfs were represented by the South American genus *Apistogramma* or the remarkably similar African genus *Nanochromis*.

The major exceptions were the African and Asian examples mentioned above, the kribs, genus *Pelvicachromis*, and the chromide, *Etroplus maculatus*, which most considered to be dwarfs, plus the African jewel cichlid, *Hemichromis bimaculatus*, which, while not large at all, had the same nasty temperament as many of its bigger New World cousins and seemed appropriately grouped with them.

Since the *Apistogramma* and their African counterparts the *Nanochromis* were relatively meek, non-destructive, cave-spawning, and rather small, and since most of the others were pugnacious, plant-and-gravel-spewing, open substrate spawning, and rather large, the idea of "dwarf" was linked to all those former traits. In addition, though some large

Nanochromis parilus is a slight challenge for the hopeful dwarf cichlid breeder. This is an African fish. Photo by Edward Taylor.

cichlids are easily sexable, some are not, but almost all the "dwarfs" were very sexually dimorphic, with males and females differing greatly in both size and coloration.

But that was before the great influx of the African rift lake cichlids of the last four decades, and to complicate matters, not only did we have this African invasion, which brought hundreds of new and often unusual species of cichlids into the hobby, we also had an upsurge in interest in *all* cichlids, with the result that today many more neotropical species are familiar and available as well. Species which were rarely or never seen in aquarium shops are now fairly easy to obtain.

This diversification of available cichlid species has shown the

original dichotomy to be of limited usefulness, like the old biological taxonomy which had just two kingdoms: plant and animal. Faced with an increasing knowledge of species which are neither plants nor animals, many but not all of which are microscopic, scientists had to add kingdoms to the classification system. There were diehards who tried to shoehorn every known life form into either the plant or the animal kingdom, but such devotion to arbitrary divisions merely hinders the progress of taxonomy.

The formation of new taxa such as Protista and Monera does not, however, affect a gardener's appreciation of certain plants or a zookeeper's understanding of certain animal species. Likewise, the expansion of the cichlid-

Nanochromis parilus is digging a pit under a coconut half-shell. She will spawn in this cave. Photo by Hans Joachim Richter.

A male *Apistogramma cacatuoides*. Photo by MP&C Piednoir.

keeping hobby does not have to detract from our appreciation of dwarf cichlids, even if it is no longer so useful a concept for dichotomizing all the species. In fact, as we shall discuss here, we can broaden the original concept to extend the enjoyment of smaller cichlids beyond anything available to earlier tropical fish hobbyists. People who are interested in the smallest cichlids do not have to worry whether their favorite species meets some set of criteria or fits into one particular group of fishes; they can simply enjoy their focus on one portion of the extreme diversity found in the family Cichlidae.

As the criteria for defining dwarf cichlids are subjective, the reasons people have for keeping and raising them are varied. Some want fish they can keep in small tanks, some want to watch the fascinating parental devotion of many dwarf species, some enjoy the "community fish" nature of many smaller cichlids, and others just "click" with these interesting and beautiful gems. Whatever your reasons for choosing dwarf cichlids, you will find the miniature end of the cichlid spectrum a rewarding place on which to focus your interest in tropical fish.

For our purposes here, we will consider two groups of fishes: (1) the traditional dwarfs such as the genera *Apistogramma* and *Nanochromis*, and (2) other small Old and New World cichlids. But first we'll review some points of basic husbandry.

Apistogramma agassizi male developing breeding coloration. This fish is from South America. Photo by Hans Joachim Richter.

EQUIPMENT

THE TANK

While many large cichlid breeders are minimalists in setting up their tanks, using bare tanks with just a piece of slate or a length of PVC pipe as a spawning site, the appeal of dwarf cichlids is, for many breeders, the fact that most of them can be maintained in fully decorated aquaria. It is even common to find them regularly spawning in a community tank.

In fact, many dwarf cichlids are timid and need both the security of a well-planted tank and the assurance of a school of small "dither fish" to coax them out of hiding, while the breeders of more aggressive large cichlids use such "target fish" to divert the belligerent tendencies of the pair away from each other. Any digging these dwarfs might do is usually limited to clearing the substrate out of their spawning cave, so plants are safe, and your little cockatoo cichlid isn't going to break the heater or pick up a rock and heave it through the tank glass the way an oscar might. And while these fish are fanatic parents, they are content to live and let live, so other inhabitants of their aquarium are also safe as long as they stay away from the dwarf cichlids' brood.

It *is* possible to breed some species in five-gallon aquaria, but it's not a recipe for success, and a ten-gallon tank is a really the

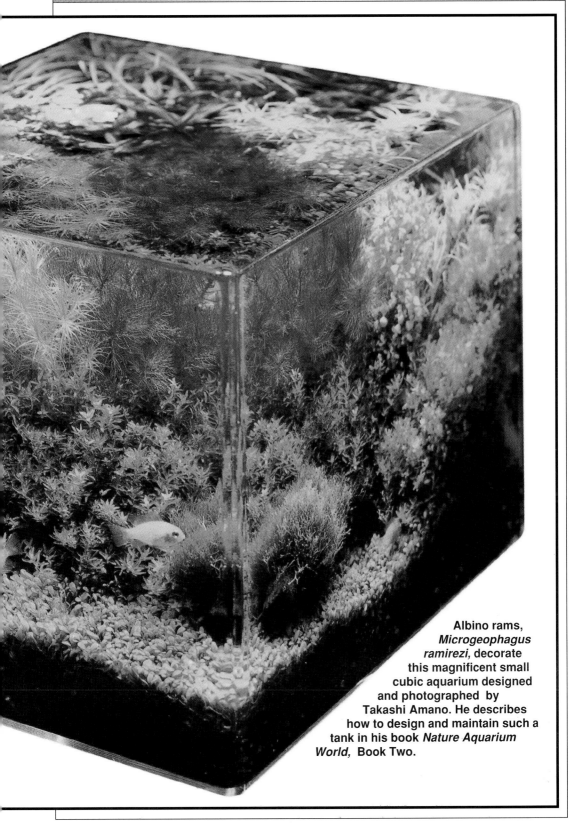

Albino rams,
*Microgeophagus
ramirezi,* decorate
this magnificent small
cubic aquarium designed
and photographed by
Takashi Amano. He describes
how to design and maintain such a
tank in his book *Nature Aquarium
World,* Book Two.

Takashi Amano is credited with being the world's best designer of aquariums and aquarium systems. Here is an aquarium designed with rocks and plants which are not easily disturbed by small cichlids. Photo by Takashi Amano.

minimum size for rearing a batch of fry, with or without their parents, and as always in caring for growing fish, bigger is better. In addition, larger tanks enable you either to breed the cichlids in a community setting or to keep a breeding group in a single-species tank.

The fact that dwarf cichlids easily *fit* into small aquaria does not mean they should be kept in them. Many of them breed as harem polygynists, with the male defending a large territory in which several females maintain subterritories. A single pair in a small tank may well breed, but if the male is ready to spawn and the female isn't, she may be battered mercilessly with no way to escape the male's attacks. Conversely, once a pair spawns, the female will drive the male away from what should be her subterritory. If the male has no larger territory to escape into, *he* may be battered mercilessly.

More than one pair of small, peaceful cichlids can have sufficient room for separate territories in a tank 36 to 48 inches long, and many species will form stable polygynous breeding populations of one male and several females in such a setting, which will enable you to observe their fascinating and complex natural reproductive behavior.

You can even use jumbo tanks for your dwarfs. The nature of these cichlids makes it possible to set up a delightful and unusual community aquarium using tanks such as the popular six-foot

It's crucial for the health of your fish that they get proper light - not just an amount of light, but a certain quality of light, too. Fortunately, there are many light systems available for fish hobbyists at pet supply stores. Photo courtesy of Energy Savers Unlimited, Inc.

aquaria—the 100 gallon or the 135 gallon. The large bottom area (nine square feet) of these tanks permits a lot of territorial manipulation and the chance for the establishment of peaceful breeding populations. It is even possible to keep many individuals of different species, together with other types of small fishes, in one large community aquarium. In addition, the 100 gallon tank's relatively short height of eighteen inches makes it possible to grow a lush forest of plants without the necessity of using metal halide lighting, since one or two double fluorescent light fixtures will suffice.

There is a growing interest in biotype aquaria, and dwarf cichlids provide an opportunity to establish such a tank on small or large scale. A South American riverine biotype with pacus, freshwater stingrays, and red-tailed catfish would be magnificent (you could use oscars as feeder fish!), and it might fit in an Olympic swimming pool, but an even greater diversity of dwarf species could easily be maintained in a moderately-sized aquarium.

Whatever size, make sure that your tank is flat and level on a stand made for that size tank, and make sure, if your tank is 30 gallons or larger, that the

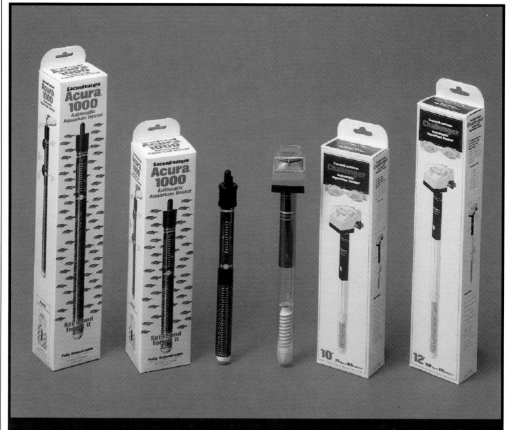

Aquarium heaters are available in a variety of sizes, wattages and features to match any hobbyist's needs. Some hang on the back or side of the aquarium while others are fully submersible in water. An on/off indicator light, suction cups and a temperature setting indicator are desirable features. Photo courtesy of Tetra/Second Nature.

floor is suitably strong. Pick a location away from temperature extremes such as radiators or outside doorways, and don't place it near a window.

OTHER EQUIPMENT

You will need a close-fitting top. A good cover not only keeps the fish in, and dust, little fingers, and the family cat out, but it also cuts down on evaporation, which, in turn, keeps the water chemistry more stable. If you are going to have a planted tank, ascertain that the light fixture is suitable for the setup you plan.

An appropriately-sized heater is necessary, of course. Many dwarf cichlids prefer the warmer end of the tropical range, in the low to mid-eighties, especially for breeding, so don't select a heater only marginally adequate for your size tank.

You will also need a means to change water and clean detritus off the bottom of the tank. This can be a simple siphon hose and a bucket, a commercial water changer/gravel cleaner, or a state-of-the-art, continuous water changing system, complete with power filter vacuuming of the gravel surface. Preference and pocketbook will direct your choice. You should think of regular water changes, however you do them, as your primary means of controlling water quality, with filtration to supplement it.

FILTRATION

Dwarf cichlids produce much less waste than their larger relatives, so monstrous power filters aren't required. High quality water *is*, however, and biological filtration is important to keep dissolved metabolites to a minimum.

In contrast to mechanical filtration, which simply strains particles out of the water with some fibrous material, biological filtration takes place in a medium with enormous surface area (fibers, sponges, ceramic rings, plastic balls, gravel, etc.) for colonization by beneficial bacteria. These aerobic bacteria convert deadly ammonia to less deadly nitrites, and nitrites into relatively harmless nitrates, which are removed by water changes.

Sponge filters will provide both mechanical and biological filtration in smaller tanks, and serve well as additional biological filtration in larger tanks which also have a power mechanical filter. In breeding tanks you must be concerned about the size of the strainer openings on the intake tubes, since you want to keep the fry in the tank, not in the filter box or canister. Sponge filters are excellent here.

All types of power filters are suitable for dwarf cichlids, as long as the water return is not so powerful that it blows the little fellows all over the tank. Many power filters have features which combine mechanical and chemical filtration with biological filtration, either through high surface area filter media, trickle returns, or rotating wet-dry wheels. You cannot put too much emphasis on this type of water purification. Undergravel filters, hailed by some aquarists and shunned by others, provide good biological filtration, and if you're one of the aquarists who likes them, you will appreciate that most dwarfs will not excavate down to the bare plate the way larger cichlids invariably do.

Many dwarfs, however, hail from soft, acid waters, and for some species these conditions must be replicated for breeding success. In such water, the biofilter is much less effective, and by the time you reach an extremely low pH, the useful bacteria cannot thrive, and biological filtration is no longer a viable option, so it has to be compensated for with more chemical filtration and increased water changes.

There is a wide variety of chemical filter media on the market, from various grades of activated carbon to resins for removing specific waste products. These can improve water quality considerably, and the extent of their use depends on the type, size, and number of fish you have in a given tank. A reliable dealer can assist you in deciding.

But no matter what type of water you maintain, and no matter how much biological and chemical

Pseudotropheus zebra, an African cichlid which just barely qualifies as a dwarf cichlid, in an appropriate aquarium setting. Photo by Andre Roth.

Rams and checkerboard cichlids thrive in this magnificent aquarium setup by Takashi Amano. To learn how to make an aquarium like this read the three volumes of his *Nature Aquarium World* series. Photo by Takashi Amano.

filtration you have, frequent partial water changes are mandatory, either to remove nitrates—the end-product of biological filtration—or to remove the types of nitrogenous wastes which will accumulate in the absence of a flourishing biological filter. Siphon-clean the gravel while drawing off twenty percent or more of the water, then replace with water of the same temperature and chemistry as the water in the tank, conditioned to remove chlorine compounds if necessary. Do this at least once a week. Water quality is of the utmost importance.

The zealous hobbyist who performs ten percent or more water changes as frequently as every day (but certainly no less than once a week) will be rewarded with healthier fish, more successful spawnings, and sparkling clean tanks. I know of a commercial breeder who produces thousands of top-quality angelfish and discus in hard, alkaline water. It isn't that he has developed strains which are unaffected by conditions far removed from their Amazonian origins, since he has the same success with wild-caught fish. To what does he ascribe his success? He changes one-third to one-half of the water in his aquaria *every single day.* Remember to think of water changes as the most important "filtration" you can have for your fish.

Most power filters provide substantial water turnover, producing good aeration. The oxygenation which wet-dry filters provide for their aerobic bacterial colonies does an excellent job of aerating the water for the fish as well. Additional aeration from airstones or air-driven ornaments is excellent, however, and can play a major or minor role in a dynamic decoration of your tank.

A pair of *Nannacara anomala* (female in foreground), tending their eggs, which were laid on the side of a cracked rock. This species normally spawns in a cave-like structure, but if a cave is not available will use a different site even one in open water. Photo by H. J. Richter.

AQUASCAPING

The substrate in your aquarium should be a couple of inches of fine gravel. Besides providing an aesthetic concealment of the tank bottom, it makes the fish feel more secure than the unnatural transparency of a glass "riverbed," and it provides a medium for anchoring your plants.

Most dwarfs prefer a cave for spawning. This can be a hole excavated by them under a rock overhang, a clay flowerpot with a notch in it which you provide, a cavity in a rock formation, a suitably-sized piece of PVC pipe or

A section of a small dwarf cichlid aquarium designed by Takashi Amano. Photo by Takashi Amano.

a PVC elbow, or a cavern in a hunk of driftwood. Some species will accept a broad plant leaf, such as of a swordplant, as a spawning site if it is in a tangle of lush plants which provide a sense of security.

For many shy dwarf cichlids, a heavily planted tank is the best environ-ment, not only for breeding but also for your enjoyment. Paradoxically, the more hiding places the fish have, the more you will see them. In a bare tank, they will be cowering behind filter tubes or vertically in the corners, and the stress will soon cause them to succumb to disease. When provided with plant thickets, driftwood tangles, and rock caves, the fish will establish territories and patrol them openly, secure in the knowledge that at any perceived danger they can dash for cover.

The choice of plant species is up to you, provided you avoid finicky species which won't thrive in the water you will be maintaining in your aquarium. For the most part, your dwarfs will not bother the plants, either by digging them up or by grazing them to stubble, the way many larger cichlids would.

Make sure any driftwood or rocks you use are safe for aquarium use, and if you build cave structures out of rocks (flat pieces of shale are excellent for

A truly magnificent dwarf cichlid aquarium featuring golden rams. This tank was designed by Takashi Amano and details of its design and maintenance can be found in his books called *Nature Aquarium World*. He also has special articles in *Tropical Fish Hobbyist* magazine. Photo by Takashi Amano.

this), fasten them to each other with silicone to make stable formations.

Your imagination is the limit for decorating your aquarium. The only items you must avoid are rocks and shells which are suitable only for marine (or African rift lake) tanks, since these can adversely affect your water chemistry. The use of plastic plants and other ornaments is a matter of personal preference. While many aquarists avoid any unnatural-looking objects, the fish will live quite happily with "No Fishing" signs, toxic waste barrels, mermaid castles, and divers swimming through the tank sides. In fact, many popular ornaments made to look like castles or sunken ruins provide excellent hiding and spawning places.

The best approach is to place plants and orna–mentation in the rear of the tank in a horseshoe shape from front corner to front corner, leaving an open swimming area in the front center. In a breeding tank you can place a spawning cave in the center of this area, which will anchor your pair's territory and provide a much better view of the proceedings than if they're hidden in the back behind a bunch of plants. Of course, such a hidden spawning site would be ideal for a timid pair you are having difficulty getting to feel safe enough to breed.

A community tank with tetras and dwarf cichlids requires fishes which neither eat plants nor dig them up! This is a Takashi Amano-designed aquarium. Photo by Takashi Amano.

WATER

You may get tired of hearing about water changes, but your fish will never get tired of having them! Think for a minute about your scaly charges. In their natural river environment, the water is constantly flowing past them. As they excrete waste or respire carbon dioxide out their gills, these noxious substances are swept away in the current. In a stalled crowded elevator, even if everyone has showered and no one had garlic with dinner last night, the air rapidly deteriorates. It's no different for the fish confined in their little glass box.

Is water "freshness" your only concern? No. The *chemistry* of your water, mainly its hardness and its pH, are also important. If you are not familiar with testing your water chemistry, get some quality test kits from a dealer who can explain them to you. Many dealers will test your water for you originally, so you can compare your results with theirs.

You need to measure two types of hardness: general hardness, which refers to the total dissolved minerals, and carbonate hardness, which indicates the buffering capacity of the water, that is, its ability to remain at a stable pH. The more the buffering capacity, the more acid the water can absorb before changing pH. Soft water has few dissolved minerals and little or no buffering capacity, while very hard water has high concentrations of dissolved minerals and substantially higher buffering capacity. Most dwarfs prefer softer water, though not all like the extremely soft reverse osmosis water made famous by discus breeders. Many will thrive in moderately hard water, and some will even breed in it. But what do you do if your water is too hard?

Soft water from your home softener has had the calcium ions replaced with sodium ions and is not really any "softer" from a fish's point of view; that is, the total concentration of minerals is the same. It is soft because of the fact that the calcium ions in hard water precipitate with soap molecules and the sodium ions in softened water do not, so with softened water you do not build up a scale on bathroom fixtures. From that point of view, the sodium ions don't count, so the water is soft.

Your fish, however, are not concerned with a soap ring in the tub or lime deposits in the toilet bowl. To soften *their* water supply, you need to remove all the ions, including sodium. You can use distilled or reverse osmosis water, or water demineralized in a resin column. You can mix a portion of this treated water with your regular tap water until it tests out properly.

Keep in mind that any adjustments you make to the

Apistogramma nijsseni, one of the less commonly available dwarf cichlid species. A male of the species is shown; a breeding - colors female would retain the red edge to the tail fin but would have a basically yellow body bearing three large black blotches. Photo by MP&C Piednoir.

water for the initial filling of your tank must be made to all water you add to the tank during water changes. Do not rely on a formula, since water supplies can vary quite a bit. Test the water, make adjustments, and test again.

As an example, there are several species of *Apistogramma* which do not breed successfully except in soft, acid water. Even if they do spawn, the eggs fail to develop properly. Some aquarists also report that in hard water the sex ratio in their spawns is unbalanced. On the other hand, others report that sometimes species showing good hatch rates only in soft, acid water also show unbalanced sex ratios; harder water gives poorer hatches but more balanced sex ratios. What

the mechanisms for these phenomena are is unclear.

In any case, it may be necessary for some species to use R.O. or distilled water with peat filtration, but remember, many fish species, especially those which have been captive bred for many generations, will do well in water which is not exactly the same as that of their ancestors' origin. Before you spend a lot of time and money altering the chemistry of your water, unless it is extremely unsuitable, you can try breeding your dwarfs. If they fail to breed successfully, or if you get unbalanced sex ratios in the young, then you can make adjustments.

As far as carbonate hardness is concerned, it is important to remember that soft, acid water is lacking in this buffering capacity, which means it is much more susceptible to extreme pH lowering, which can stress and even kill your fish. Think of carbonate hardness as a pool of buffers, just waiting to join up with free acid (H⁺) ions. As natural metabolic and decomposition processes produce acid, the buffers combine with the acid to neutralize it. This situation, by the way, accounts for why hobbyists who use pH lowering products in very hard water get frustrated; the highly-buffered water simply soaks up the added acid, and the pH quickly goes back up, necessitating constant, repeated treatments. In soft water, on the other hand, any acid produced in the tank simply builds up, often with alarming speed. Low buffering capacity thus poses a serious threat in a tank containing soft water. Even in a well-established and flourishing aquarium, a significant increase in acid production, say from an

overfeeding or a dead fish, or even an increase in temperature on a hot day, which results in additional carbon dioxide (and hence acid) in the water, can cause a rapid crash of the pH. The disaster only accelerates even more as the anaerobic bacteria start to succumb to the acidic environment, putting even more stress on the fish from ammonia poisoning.

Laetacara dorsigera does best in soft, slightly acid water. Photo by MP&C Piednoir Aqua Press.

So, besides being important in itself, water hardness can be important with regard to the other important measurement you must make—pH. The pH (proportion of hydrogen ion) of water is measured on a scale of 0 to 14 and gives the concentration of H⁺ ions (acidity) in the water. Neutral water, such as distilled water, has a pH of 7.0, with the scale normally being from strong acid at a pH of 0.0 to strong base (alkaline) at a pH of 14.0. This is an inverse logarithmic scale, meaning that a change from a pH of 6.0 to 7.0 is a *tenfold decrease* in acidity.

This *Melanochromis* from Lake Malawi, Africa requires very hard water. All cichlids from the Great Rift Lakes of Africa require hard water. Photo by Edward Taylor.

Do do not minimize what might look like a small difference in pH. For example, your fish will not notice the difference between 78°F and 78.5°F; most thermometers cannot even measure such small differences, and the fish might be comfortable in a ten degree range from 72° to 82°F. A difference of 0.5 pH, however, is a more significant deviation; it is easily measured, and very few species could tolerate a pH range of even four points, say, 5.0 to 9.0. Most freshwater aquarium fish prefer a pH somewhere between 6.5 and 7.5, and while many dwarfs come from more acidic natural environments, some of them come from neutral to slightly alkaline waters. Due to the variability of habitats, it is impossible to give an "ideal" pH even for the single genus *Apistogramma*, let alone for, say, all South American dwarfs.

We've already mentioned the effect of pH on biological filtration, but it bears repeating. Your regular water changes must be greatly increased in acid water tanks, since you cannot count on an effective biofilter at a low pH.

Besides water chemistry, you must monitor the temperature. Very few dwarfs like temperatures as cool as the low seventies, and most require eighty or better for breeding. You should adjust your tank's temperature to match the requirements of the particular species you wish to keep.

FOODS AND FEEDING

There is no mystery to the effective feeding of tropical fish, and dwarf cichlids are no exception. High quality food, in great variety, fed sparingly many times a day, will produce the healthiest fish and the most breeding successes. There are great debates about live versus frozen foods, freeze-dried versus prepared foods, etc., but the bottom line is this: feed high quality food, in great variety, sparingly many times a day.

When feeding fry, it's the same situation. Whether you feed live brine shrimp, powdered flake food, homemade infusions, or anything else of suitable size, you must feed tiny amounts very often to get maximum growth.

Fortunately most of the dwarfs discussed in this book will readily take all of the common aquarium foods, and the choice is up to you. A few species, especially if wild-caught, are a bit finicky and eat better when fed frozen or live foods, though there is a lot of individual variation, plus an increase in adaptability in successive generations raised in captivity. Because all fish relish them so much, though, live, frozen, and freeze-dried food organisms quickly bring fish into breeding condition, simply because they voraciously eat these highly nutritive foods. With the plethora of excellent, scientifically formulated alternatives available on the market today, however, no aquarist is forced to feed any particular type of food, and it is possible to keep and even breed fish without any live food.

A typical daily feeding program might be: one meal of frozen brine shrimp, one of freeze-dried bloodworms, one of a flake food, and one of a small-sized pelleted food. If you are not home during the day you may not be able to feed four times a day, so make it two or three slightly larger feedings. If you are raising fry and can manage six or more feedings a day, their growth will astound you. Of course, there are available a number of automatic feeders that can be filled with a variety of foods (not live or frozen, of course) and programmed to feed any number of times a day at various intervals.

Even if you were to feed only once a day, if you divide that feeding into several portions and feed them out over a period of a half hour to an hour, you will see less waste, less strain on the tank's filtration system and on the fish, and healthier fish.

The vast majority of smaller cichlids are either predators, micropredators, or omnivores, so feeding vegetables should not be a major concern. Commercial foods contain all the vegetable matter dwarf cichlids need, and if they really have a hankering for a salad, they can graze on a little algae. I have found that the specially formulated cichlid diets, which are designed to be adequate for both African rift lake and neotropical species, satisfy omnivorous species quite well, though they seem to leave the African species which are vegetarian looking for more greens, perhaps because eating such prepared foods is such a different behavior from the browsing and grazing by which those fish naturally get their meals.

organisms or with chemical pollutants. Brine shrimp, of course, since you hatch them out yourself, are safe, and are the universally preferred first food for fry; they are even suitable for the adults of many dwarf species. Another excellent live food with very little chance of introducing pathogens to your fish is the common fruit fly, *Drosophila melanogaster*. It is available in a wingless mutation, the escapees of which won't wind up buzzing around your fruit bowl. The flies are easily cultured, easily fed, and

Fish need protein and other nutrients in their food, and they get more out of their food when the quality of the ingredients is high. Photo courtesy of Ocean Star International.

Recently live foods have fallen into considerable disuse, not only because of an avoidance of the mess and fuss of live foods in light of the excellent alternatives available, but also because of the possibility of the live foods being tainted, either with disease

enthusiastically eaten by fish (they float on the surface—but not for long!) Though not usually available in pet shops, they can be mail ordered. They can be cultured on mashed bananas, but mold, always a threat in *Drosophila* cultures, can be quite a problem, and you really have a mess when it comes to dumping flies out of the container. Powdered commercial or cooked homemade media with mold inhibitors are much more practical, and supplies and

contaminants such as heavy metals. There is no longer any necessity in feeding these worms, though fish do like them, and the freeze-dried ones are safe.

My fish have always preferred freeze-dried bloodworms (which

Tubifex worms are ideal foods for fishes if the worms are cleaned. Unfortunately it is very difficult to clean them. DONT HANDLE THEM WITH YOUR BARE HANDS AS THEY MAY CARRY HARMFUL ORGANISMS. Photo by Mark Smith.

culturing instructions are available from the suppliers of the flies.

Live *Tubifex* worms, once a staple for aquarium fish, are very prone to contamination, given their natural habitat of sewage drainage areas. They could conceivably even make *you* sick! At least one company offers pathogen-free freeze-dried tubifex, which are irradiated to kill all disease organisms, though they could still contain other

also are not without their opponents because of potential pollutant contamination) to freeze-dried tubifex. The bloodworms also have the advantage, shared with freeze-dried brine shrimp, of being able to be easily pulverized between thumb and forefinger to make a powder suitable for feeding newly free-swimming fry, something you cannot do with a cube of freeze-dried tubifex.

Almost all cichlid fry can handle brine shrimp nauplii from the first, and millions of tropical fish owe their existence to this wonderful food. An admittedly much smaller number of fish, however, are raised to healthy adulthood never having seen a live brine shrimp. If you are not able or inclined to hatch brine shrimp, you can still raise your dwarf cichlids on such alternatives as the powdered freeze-dried worms and shrimp mentioned above. (There is always

small size of pellets now available. These tiniest pellets, designed for tetra-sized fish, are suitable for most cichlid fry not too long after they hatch. The old rule of thumb to feed fish food no larger than their eyes will give you a rough idea of when your babies can handle these pellets. Try a few, and if they are consumed easily by the fry, you can start including them in the diet. There is, of course, always some discrepancy

A female *Nannacara anomala* surrounded by a cloud of her fry. The fry shown here are mostly uniform in size, buy many times cichlid fry grow at a greatly varying rates. Photo by H. J. Richter.

a ready-made supply of perfect fry food at the bottom of a container of freeze-dried shrimp.) Flake foods can be powdered in the same manner, and they are also sold already pulverized for fry.

Another commercial product which has greatly simplified the raising of cichlid fry is the very

in size among any cichlid hatch, and males often grow faster than females. It is interesting to note that when only the largest of a clutch are able to handle these "micro" pellets, their smaller siblings attack the pellets avidly, pecking away at them. This food is practically waste-free, since it is

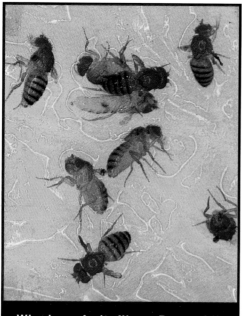

Wingless fruit flies, *Drosophila melanogaster,* can be raised in your own home very easily. Obviously they cant fly. Photo by Michael Gilroy.

eaten long before it softens and falls apart in the water. It makes an excellent part of a varied diet for growing youngsters.

Whatever you choose to feed your cichlids, remember that it is not necessary to fuss about their food; you can feed them well and simply. For many hobbyists, however, culturing various food organisms and preparing homemade concoctions is part of their enjoyment in providing their pets with the best possible care. There is nothing inherently preferable in such a regimen, or in a simpler one of using only commercial preparations. Today as never before the aquarist has a multitude of choices in food items to fit his or her preferences, budget, and needs.

Mosquito larvae ae excellent as a fish food but if the fish miss them, the adults might fly away and search for blood.! Photo by R. Scheriber.

The larvae of the chironomid fly is known as a bloodworm by aquarists and is one of the preferred live foods. Photo by Dieter Untergasser.

THE TRADITIONAL DWARFS

The traditional dwarf cichlids are of the two genera already mentioned, *Apistogramma* and *Nanochromis*, plus several similar genera. These bantam species of African and neotropical cichlids come from streams, ponds, and rivers filled with vegetation and supplied by rain runoff. Due to their small size, they feel tremendous pressure from predation, so they have retiring personalities, preferring plant thickets, the roots of floating plants, driftwood tangles, and sunken leaf litter as habitats. The water is often but not always soft and acid. The shallow habitats they frequent normally have a relatively high temperature.

Fish of these genera typically show sexual dimorphism; that is, the males and females are quite distinct in appearance, with the males being considerably larger. *Apistogramma* males are typically more colorful than their mates; among the *Nanochromis* this is also usually the case, but in the appropriately named *N. transvestitus*, as among the kribs, the females' (belly) colors outdo the males'.

A female *Apistogramma borelli.* Photo by Klaus Paysan.

A male *Apistogramma borelli* in breeding color. Photo by Hans Joachim Richter.

Apistogramma

The hardest part in obtaining species of this venerable South American genus might be finding out what you've found. There are more than fifty species in the genus, and it is not uncommon to see for sale a tank of mixed dwarf cichlids which contains various *Apistogramma* species. It may not even be possible to tell them apart so as to get a male and female of the same species. Add some confusion in the literature about their taxonomy, and you have a truly perplexing situation. If at all possible, find a dealer who has single-species named supplies and keeps them separate.

They all breed in the same way, spawning in a cave, after which the female drives the male out and tends the eggs while he patrols the immediate territory and breeds with other females in the tank. In a few species the male is tolerated around the young after they have grown a bit, though it is always a good idea to remove the male after spawning if you are keeping them as pairs, for the protection of both the male and the fry.

Three well known species are *A. agassizii*, *A. borelli*, and *A. cacatuoides*, though there is a historical confusion between the last two, as there is between *A. cacatuoides* and *A. jujurensis*.

A. agassizii is first not only alphabetically. Known to science for more than a century, it is the "type species" for the notion of dwarf cichlid, and its attributes are the ones traditionally generalized to all dwarfs. For it is small (males about three inches, females less than two inches), comes from South American rivers, is very dimorphic, undergoes considerable color change during breeding, likes warm, soft, acid water, is peaceful, needs a planted tank and does not bother the plants, spawns in harems in caves, with the females caring for the fry, etc.

Though certainly not gaudy or brilliant in coloration, this fish is quite pretty and is available in several color morphs of red, yellow, and blue, with the male much more colorful than his mate. It has a distinct dark lateral band from the eye into the caudal fin. It lacks the extended dorsal rays of many dwarfs, but shows typical sexual dimorphism. Aside from differences in size and color, the tail of the male is

A pair of *Apistobramma agassizii* spawning. The male is the more colorful of the pair. Photo by Hans Joachim Richter.

A pair (male above) of one of the aquarium-cultivated color varieties of the Congo cichlid, a species that is perhaps the easiest of all cichlids—dwarf and non-dwarf—in which to induce spawnings. Photo by H. J. Richter.

lanceolate (spearpoint shaped), while that of the female is rounded.

For breeding a pH of 5.0-6.0 is best, with a temperature of 80°F. Spawning is typical for the species. Some breeders prefer microorganisms ("infusoria") for the first few feedings, and others start right in with brine shrimp nauplii. In an established tank with a sponge filter, there probably are enough infursorians to feed the young at first, but it is a simple matter to know if fry are big enough to consume the shrimp—their bellies turn fat and orange.

A. borelli, slightly smaller than *A. agassizii*, has a higher, less elongate body and long anal and ventral fins. The lateral band is normally restricted to the caudal fin and peduncle. This species also has several color morphs. Soft to moderately hard water, neutral to slightly alkaline (ph 7.3) is ideal, with temperatures around 80°F for spawning.

Peaceful even for the genus, this species is ideal for setting up several pairs in a large aquarium. The female is unusually tolerant of the male, who may join her in brood care.

A lovely male *Apistogramma steindachneri*, an ideal dwarf cichlid if you can find them at your local pet shop. They are rare. Photo by Aaron Norman.

A. cacatuoides, the cockatoo cichlid, gets its name from the cockatoo "crest" formed by the elongated first rays of the male's dorsal fin. The ventral fins are short, with one elongated ray. Of the many color varieties available, the most striking is the double red, in which the male has large red areas, bordered in black, in the lyre-shaped caudal fin. An orange variety also exists. Both sexes have the typical dark lateral band, plus several vertical bars.

Another dwarf that bucks the soft-acid stereotype, this species can be found in moderately hard water, with one reported location having a hardness of up to 14° and a pH of 7.6! It is especially inclined to harem breeding, so the ideal setup is one male and several females in a sufficiently large aquarium.

The rest of the species in the genus follow suit, and even if you aren't sure which one you have, you can enjoy keeping and breeding them. There exist several excellent books with color photographs which can assist you in identifying individual species and their precise preferences in water chemistry.

The cockatoo cichlid, *Apistogramma cacatuoides,* was named for a cockatoo after the male was observed with his extended first five dorsal fin rays. Photo by Hans Joachim Richter.

Apistogramma trifasciatum. This male looks like *cacatuoides* except for the color. Photo by Hans Joachim Richter.

Apistogramma wickleri male. Photo by Klaus Paysan.

Nanochromis

Africa and South America are the realm of the family Cichlidae, reflecting their one-time connection to each other. Continental drift brought us the South Atlantic Ocean, and genetic drift and other evolutionary factors produced an amazing variety of cichlids on these two continents. The niche filled by the South American dwarfs is, in Africa, filled mainly by fish of the genus *Nanochromis*.

This genus, whose name means "dwarf cichlid," offers about ten species, almost all of which are under three inches, with females often about two inches. These Africans are a bit rarer in the hobby than the Apistos, but several are usually available. Many are quite beautiful but are overlooked because their non-breeding colors are rather drab. They all need well-concealed caves or cavities in which to breed, where they hang their eggs on the ceiling. Mother and father both care for both eggs and fry and confine their aggressions to the immediate area of the cave, so even when breeding they make good community fish. You may have to hunt a bit for these fish, but larger stores should have one or more species, such as *N. dimidiatus*.

Nanochromis parilus used to be called *Nanochromis nudiceps*. Photo by Edward Taylor.

Nanochromis dimidiatus, a male.

THE SMILING ACARAS

The genus *Laetacara* contains several wonderful aquarium species. Their name refers to the dark line which "paints" a smile on their mouth. They are miniature acaras, with most species three inches or less.

The species *L. curviceps* is a perennial aquarium favorite, known for years as *Aequidens curviceps*. This delightful cichlid is a true dwarf acara; it looks like a miniature port cichlid, with lovely iridescent spangling on the sides. They appreciate a well-planted tank, but they are curious and cautiously outgoing—great inhabitants for a community tank. They are about

A pair of *Laetacara curviceps* spawning. The male (lower fish) is fertilizing the eggs. Photo by Hans Joachim Richter.

Laetacara curviceps. This is a male in preliminary breeding colons. Photo by Hans Joachim Richter.

Laetacara thayeri collected and photographed by Dr. Herbert R. Axelrod in the Brazilian jungle. These are valuable photos because most of the beautiful photos are taken in home aquariums where the origin of the fish is usually unknown.

three inches, with the female slightly smaller than her mate.

This fish is not too fussy about water, though soft and slightly acid is best. They spawn just like the larger acaras, on some flat surface, preferably concealed behind plants, and both parents care for the eggs and young, moving the babies repeatedly from pit to pit until they become free-swimming.

Equally recommendable and even more colorful is *L. dorsigera.* While a drab black on grey when immature, when in breeding colors both sexes have a purplish-red wash over the throat and belly, where *L. curviceps* tends to bluish. Care is the same as for *L. curviceps.* Larger and slightly less colorful is *L. thayeri,* which, at about four inches, is almost one of the in-between-sized cichlids.

Laetacara dorsigera, a pair with their free-swimming fry. Photo by Jaroslav Elias.

OTHER SOUTH AMERICAN DWARFS

A perennial favorite among South American dwarf cichlids is the ram, *Microgeophagus ramirezi*, which you will see in older literature as *Apistogramma ramirezi*, and in some newer literature as *Papiliochromis ramirezi*. The larger *M. altispinosa* has had periodic popularity and is occasionally offered, but the ram is almost ubiquitously available. The beauty and charm of this species has ensured its continued

Available in both its normal wild-color form (as shown) and a golden color variety, the ram is a perennial dwarf cichlid favorite. Photo by Hans Joachim Richter.

Microgeophagus or *Papiliochromis altispinosa* is closely related to the ram. Photo by MP&C Piednoir.

Two *Microgeophagus ramirezi* squaring off against each other. Photo by MP&C Piednoir Aqua Press.

popularity, even though it has a reputation for being hard to keep and breed. This is at least partly due to the characteristics of its natural habitat, with a hardness under 1°, a pH of about 5.0, and temperatures near 90°F. Few aquarists are accustomed to providing such conditions.

Best kept to themselves or with a few tetra dithers, these small fish will spawn in the open on large pebbles, small rocks, or an excavated pit, producing a clutch of eggs which both parents care for in typical cichlid fashion.

The genus *Nannacara* gives us two or three species of extremely

Nannacara anomala male. These are peaceful dwarf cichlids about 2 inches long. Photo by Hans Joachim Richter.

Sexes are quite similar, though the mature male develops an elongated dorsal fin ray, and several domestic color forms and fin types have been produced. They are spawned in large numbers by Asian breeders, and these fish seem to be somewhat more tolerant of typical aquarium water chemistry.

peaceful two-inch dwarf acaras, which also hail from very soft, quite acid water. They are shy and prefer well planted tanks with plenty of hiding spaces, and caves for spawning. The female assumes all care and defense of the eggs and young. *N. anomala* has been available for many years and enjoys a steady if small popularity among specialized hobbyists.

Although not widely available, fish of the genus *Crenicara* (*Dicrossus*), the checkered or checkerboard cichlids, are occasionally offered. These are larger dwarfs, with four-inch males and two-inch females, and they require soft, acid water, less than pH 6.0. A broad leaf is the preferred spawning site. Definitely harder to keep and breed, these dwarfs are perfect for the experienced aquarist looking for more of a challenge. Spawning reports are contradictory as to whether the male assists in brood care or not.

THE OTHER TWO DWARFS

The kribs are probably, along with the ram, the most commonly seen dwarf cichlids in dealers' tanks. Various species of the genus *Pelvicachromis*, along with, unfortunately, hybrids of them, are available. The most popular for a long time has been *P. pulcher*, long known as *Pelmatochromis kribensis*, hence its common name, which was too well entrenched to change when the scientific nomenclature was revised.

Sexing these fish is easy, even at a fairly early age, as the female develops a rounded belly in contrast with the longer, streamlined male, and while

A male *Crenicara filamentosa* from Brazil. Photo by Hans Joachmi Richter.

A male *Pelvicachromis pulcher*. Photo by MP&C Piednoir Aqua Press.

coloration is extremely variable, the bright red belly of a ripe female krib is a familiar attribute of this beautiful, peaceful fish. They prefer soft and slightly acid water, but they are somewhat tolerant. They are good feeders, and live and frozen foods should bring a healthy pair into condition readily. They will spawn in a cave, traditionally an upside-down clay flowerpot with a notch knocked in the rim, and you'll know they have eggs when the male is rudely ousted from the cave. He then guards the periphery of his territory zealously. When the mother appears with her swarm of babies, the male continues guard duty.

The Asian genus *Etroplus*, a geographical oddity in the generally cichlid-free East, contains two species encountered in the trade, only one of which is a candidate for inclusion among the dwarf cichlids. *E. maculatus*, the orange chromide (also known as the yellow chromide, with a domesticated color morph known as the red chromide), is a very pretty little fish of under three inches. They spawn easily in typical cichlid fashion, generally in the open. They are quite peaceful for a cichlid, omnivorous, and normally make wonderful parents, with both sexes tending and guarding the brood. They are most comfortable with some marine salts added to their water to make it slightly brackish.

Two views of *Etroplus maculatus,* from the sub-Indian area including Sri Lanka. The lower photo shows the pair spawning with the male being the lower fish. The upper photo shows a color variety. Photos by Hans Joachim Richter.

OTHER SMALL CICHLIDS

RIFT LAKE CICHLIDS

There are many other cichlid species which share some traits with the fishes already discussed. For example, someone interested in *Apistogramma* species might enjoy keeping the smallest species from Lakes Tanganyika and Malawi in Africa. While some of these are truly diminutive cichlids, their requirement of extremely hard, extremely alkaline water puts them as dwarf cichlids (and, in fact, as cichlids in general) in a class apart. As long as you remember to provide appropriate conditions for them, however, there is no reason to exclude these species from your appreciation of dwarf cichlids.

The astounding diversity of cichlids is manifest once again in these tiny rift lake species, and it is fascinating to consider the different answers the family has come up with to the same types of challenges. In the tangles of South American jungle streams, the challenge of being so small as to be meal-sized for almost everyone else drove the dwarf cichlids into the plant thickets and leaf litter substrate. In the rift lakes, where there is little other cover, but where the competition for refuge holes among the ubiquitous rocks is fierce, a few species have opted for a secretive lifestyle and are (at least for rift lake cichlids) quite peaceful. But the vast majority of smaller rift lake species are, if anything, even more bellicose than their larger cousins.

If you wish to try the rift lake dwarfs but are unfamiliar with fish from these habitats, consult any of the excellent reference books about these fishes.

THE SHELL DWELLERS

Called "shell dwellers" because they live and breed in empty snail shells, these members of the genus *Lamprologus* are beautiful, interesting, even comical species, whose size is best described as tiny, some not exceeding one inch in length. While assertive and more combative with their own kind than some of the traditional dwarfs, they are fairly peaceful, especially toward other species. Most of them will spawn in harems, and a moderately sized tank can house a breeding group.

These are truly the rift lake dwarfs, expanding the notion of dwarf cichlid to include fish from extremely hard, alkaline waters. Otherwise they are quite like the traditional dwarfs, except they are not shy and retiring! It is a delight to see such tiny fish bullying their way to the best spots at feeding time, though obviously the aquarist should choose the tankmates of such lilliputian fish carefully.

Commonly available species include *L. multifasciatus*, *L. ocellatus*, and *L. brevis*. They require large snail shells, or small PVC elbows, not only for breeding, but as permanent housing. They will sit in the opening, watching the world go by, and dart back in if they are disturbed during an outing or if they see a usurper too close to their lair. Like many other *Lamprologus*, even the adults relish live baby brine shrimp, though they are eager feeders on any normal fish food. If you want to try some Tanganyikan dwarfs, or if you want dwarf cichlids but have hard, alkaline water, shell dwellers are for you.

OTHER TANGANYIKANS

Lake Tanganyika offers several non-shell dwelling species which are under four inches, some under two inches. The major reason these rift lake cichlids are not normally thought of as dwarfs is that even small species usually require rather large aquaria to deal with the extreme aggressiveness of these fish, either to provide sufficient territories in a single species tank or to permit the housing of individuals of many different species to spread the fighting

A shell-dwelling *Lamprologus* from Lake Tanganyika, Africa. Photo by Hans Joachim Richter.

A pair of *Lamprologus brichardi*. They move like ballet dancers! Photo by Hans Joachim Richter.

around. If small fish appeal to you, but you don't mind the larger aquarium temperament, there are many to choose from.

L. leleupi with its neon yellow color is a popular cave-spawner. The also-elongate *Julidichromis transcriptus* and *J. ornatus* are probably the easiest Tanganyikans to spawn and are quite attractive with their light and dark markings. These are also cave spawners which need a jumble of rockwork or something similar to feel secure.

Lamprologus toae, L. brichardi, and *L. pulcher* are all medium sized, and while some will grow out of the four-inch range, like other rift lake species they begin spawning at about half the adult size, so you have the option of having a tankful of "dwarfs" that will eventually outgrow their welcome.

L. caudopunctatus is a two-inch beauty which is feisty enough to hold its own against larger cichlids. Like the other members of its genus, it is almost impossible to sex, but a batch of young raised together will not only give you naturally bonded pairs, it will give you the most chance for successfully keeping the fish together in a large tank, since like most other African cichlids, these cichlids are particularly intolerant of newcomers.

Dwarf Mbuna

Lake Malawi is not without its diminutive species, though it is hard to think of them as dwarfs when they are so aggressive and require such large tanks. While many species grow to five inches or more, they begin breeding at two inches or so. At any size, however, these are all African rift lake in behavior, and they need large aquaria to prevent fatal aggression.

The same is true for the more dwarf-like *Pseudotropheus* species which are full grown at two to three inches, several of which belong to the yet unclassified "*Pseudotropheus* aggressive" complex. It is not surprising that in this habitat, where boldness and belligerence are a way of life, the smallest mbuna are among the most aggressive. What they lack in size, they make up for in fury.

Likewise, several *Melanochromis* species are in the three-inch or less range, but they are among the most pugnacious fish in the hobby, and a single male can terrorize a very large aquarium full of very large cichlids.

Cynotilapia afra males are three inches long, and the females two inches. Several *Labidichromis* are about the same size and are usually less violent than the other genera. I have had *L. chisumulae*

Lamprologus caudopunctatus. Photo by Hans Joachim Richter.

Lamprologus toae shown in its natural habitat in Lake Tanganyika. Photo by Glen Axelrod.

Neolamprologus pulcher. Photo by MP&C Piednoir Aqua Press.

Julidochromis ornatus. Photo by Edward Taylor.

Lamprologus leleupi spawning in their cave. Photo by Hans Joachim Richter.

spawn at just 1 ¹/₂ inches, and such small cichlids provide an interesting "dwarf" mouthbrooder.

Dwarf Mouthbrooders

The traditional dwarf mouthbrooders belong to the genus *Pseudocrenilabrus*, including *P. multicolor*, the Egyptian mouthbrooder, and *P. nicholsi*, the Dwarf Congo mouthbrooder, both of which do not exceed four inches. They are

Cynotilapia afra. Photo by Dr. Herbert R. Axelrod.

Pseudocrenilabrus multicolor, the Egyptian mouthbrooder. Photo by Hans Joachim Richter.

not hardwater species but come from the rivers, streams, and ponds of Africa. The males are very brilliantly colored in their breeding attire, though the females are drab. They are fairly aggressive, but one male and several females can be kept in a 36 to 48 inch aquarium with plenty of hiding places. The one to two dozen eggs are scooped up by the female as they are laid and fertilized in her mouth as she picks at them and at the anal fin of the male. She broods them for about two weeks, after which she will be thin and weakened and not up to the attentions of the male. If you have not removed her to a nursery tank for the sake of the young, you might have to remove her now for a short recuperation.

Pseudocrenilabrus multicolor, the Egyptian mouthbrooder. This is a female with a throat full of fry. She doesnt usually eat when she carries her eggs in her mouth. Photo by Hans Joachim Richter.

Small Central American Cichlids

Back in the New World, for those who want something smaller than the truly big cichlids but don't mind some of the habits of larger species, the related group of fishes containing the convict cichlid offers several under-five-inch species which are colorful, interesting, and easy breeders. Like many cichlids, they begin spawning at a very small size, so you have the option of breeding them and then trading or giving them away when they get bigger.

The easily confused species *Herichthys spilurum, H. sajica,* and *H. spinossisimus* are wonderful beginner's fish— attractive, easy spawners. Males are usually about four inches, with the females an inch or so smaller, though there are some larger and some smaller specimens. They undergo considerable color changes during breeding, make good parents, and produce lots of fry on a regular basis.

Care is the same as for the convict, *H. nigrofasciatus*, which is commonly the beginner's first cichlid spawned. This species reaches sexual maturity *extremely* early. I have had a pair of "fry" spawn when the male was only an inch long, and the female was three-fourths of an inch! They actually produced about twenty fry, which they successfully protected for almost two weeks before losing them to the other convicts and the two-five inch port cichlids in the tank.

Besides the prison-garb-striped wild type, there has been a dark-eyed white form available for decades, and recently a marbled variety has been developed, which hopefully will soon be available in stores.

These are extremely hardy and tolerant fish, and it is possible for young convicts used as feeders for predacious giants to escape capture and grow up to spawn and terrorize the fish they were

Herichthys spilurus. **Photo by Dr. Herbert R. Axelrod.**

intended to feed. They are not fussy about temperature or water chemistry, and they will eat absolutely anything. I have had them spawn in unheated tanks and in a tank located near the ceiling in a room where temperatures that high up cycled drastically through the day, ranging from the low seventies to over ninety. Obviously, it is better not to subject your fish to such extremes, but this does illustrate their outrageous hardiness. The species in this group other than the convicts are not quite so adaptable, but all of them can be cared for by keeping water chemistry and temperatures within normal tropical aquarium ranges.

All you need to spawn a pair of these is a fifteen or twenty gallon tank, or larger, with a hiding place and a spawning place. The hiding place is for the member of the pair that gets chased, usually, but not always, the female. On occasion the female will drive off the male after the fry hatch. The spawning place can be a cave, but they also will spawn on a variety of horizontal or vertical surfaces, including the tank sides.

Often you can simply buy a male and female and have success, though the best results come from the time-honored method of buying six or more juveniles and raising them together. You have to keep an eye on things and remove one if it starts getting battered. Since a pair of these is often disparate in size, you can make use of this and provide a pipe section or other hiding place small enough to exclude the male, where the female can take refuge. Usually once you get a pair on the same breeding cycle, things go smoothly.

Gravel is not necessary, though they like to dig pits into which they move the fry before they become free-swimming. For ease in cleaning I prefer bare-bottomed tanks, and the fish simply move the babies around in the corners instead. A sponge filter or two will provide safe filtration and aeration. Put one near the heater to avoid different temperature zones in the aquarium.

All of these will spawn readily in the community setting, but in most cases either you will lose the fry to predation by other fish in the tank, or you will lose the other fish in the tank from the savage attacks of the parents upon them.

A particularly delightful species is *H. centrarchus*. At about four inches, it is a stocky fish with numerous dark, vertical bars on a grey background. When the eggs are laid, the female becomes almost completely black, and the male follows suit when the fry hatch and become free swimming. The young graze on their parents' slime like discus fry in addition to feeding heartily on live brine shrimp or fine prepared foods, and the dark parents surrounded by their tiny offspring, which are light gold with a dark horizontal stripe, make a beautiful sight.

They have engaging "personalities," much like the famous oscar, and are particularly faithful and devoted parents. I once spawned a pair in a bare twenty gallon high tank into which I put two plastic "caves," which had suction cups on their bases. One I attached to the middle of the bottom, and one I put on a glass side. They spawned in the bottom cave, and when the fry hatched, the female moved them to the other cave, hanging them in bunches on the outside of the cave like tinsel on a Christmas tree, where she jealously guarded them.

Another small, popular fish is the rainbow cichlid, *Herotilapia mutispinosa*, which, though quite colorful with variations on a brown and yellow theme, is a prime contender for the most misnamed fish. It is another easy

Herotilapia multispinosus, a male. Photo by Hans Joachim Richter.

breeder which will not outgrow a twenty-gallon tank.

DWARF PIKE CICHLIDS?

The *Crenicichla* pike cichlids comprise mostly large, predatory species suitable only for the most dedicated big cichlid hobbyists.

This genus, known to science for over 150 years, has within the last decade been added to as several dwarf rheophile (rapids-dwelling) species were collected, including *C. compressiceps* and *C. regani.* The idea of a pike cichlid which can be kept in normal tanks and fed on baby guppies and live invertebrates is exciting to many cichlid specialists, but as this genus has well over fifty species and there is almost no agreement among taxonomists about them, it will be some time before dwarf pike cichlids are well understood. Some of the reported "dwarfs" may, in fact, be the juveniles of certain other *Crenicichla* species.

There have been reports of breeding these dwarf pikes, and hopefully aquarists will soon have available stocks of three to five inch versions of these large, fascinating predators, which are so intriguing but so difficult to maintain due to their size and their insistence on living prey.

Crenicichla notophthalmus, a dwarf pike cichlid, but still a predataory fish. Photo by Uwe Werner.

HEALTH CONCERNS

We've really already discussed almost everything you need to know about keeping your dwarf cichlids healthy. Frequent water changes and a varied diet will go a long way to preventing disease outbreaks among your fish. Since it is much easier to prevent disease than to treat it, and since it is rather difficult for the layman to diagnose fish illness before the fish dies, and since using aquarium medications is often a hit-or-miss affair, you should do anything you can to eliminate the stresses which can cause disease. In the event that you do have medical problems, you should seek the advice of a competent professional, but preventing stress on your fish should avoid this necessity. A properly functioning biological filter, frequent partial water changes to eliminate the accumulated nitrates, appropriate, non-fluctuating temperature, reasonable stocking rates, suitable hiding places, suitable tankmates, a tight cover, and the proper diet will prevent almost all of the stresses which aquarium fish are subject to.

The only remaining source of stress is the introduction of pathogens into a healthy system. Now, all fish, like all animals, are exposed to pathogens, since they are not raised in a sterile environment. And like all other animals, fish develop resistance to pathogens to which they are exposed gradually and in subacute concentrations.

Severe fin and tail rot in a cichlid. Photo by Dr. Herbert R. Axelrod.

The easiest way to upset this situation is to remove a fish from the environment in which it has become resistant. This, of course, is done all the time in the hobby as we buy, sell, import, export, and trade fish. It is a testimony to the immune responses of the fish that they don't all drop dead, especially when you figure in the physical and psychological stresses of netting and transport.

It is not uncommon in the fish trade for a few individuals to contract diseases during their trek, and the first rule you must follow is to avoid purchasing any fish which shows any sign of illness or stress, or which is in a tank containing any other fish which show such signs. Reputable dealers intervene on your behalf and will not sell from such a tank, and many maintain rigorous quarantine procedures, holding fish back from sale until they are certain of their good health.

The best answer to this real and significant problem is the venerable quarantine tank, that extra tank you refrain from using except when you get new fish. Now the idea of an empty tank is anathema to most hobbyists, but there is a solution to that problem as well. Have a tank—it doesn't have to be very big— set up with a sponge filter and some hiding places. Keep a few hardy fish in it to maintain the biofilter, and *resist the temptation to stock the tank to capacity!* You can still enjoy the aquarium, and should some new arrivals give a disease to the inhabitants, there are only a few inexpensive fish involved, and you haven't risked your precious collection.

There are three purposes to quarantining newly acquired fish, though many people only consider the first.

(1) By keeping new fish from your collection for at least two weeks and preferably more, any diseases which they may have contracted in their travels and which are incubating should become apparent, and you can treat them and make sure they are well before introducing them to your other tanks.

(2) By quarantining new arrivals, you separate fish which have been stressed, and stress brings out any latent health problems they normally carry in an asymptomatic state but now succumb to because of the stress.

(3) By isolating newcomers, you give them a chance to recuperate from all that stress of being caught, bought, trans– ported, and exposed to who knows how many new pathogens in the various holding tanks they have passed through. That way, when you put them into your regular tanks, they are in tip top shape, ready to handle their new environment. This purpose is helped along if after a couple of weeks you make several water changes in the quarantine tank, replacing the water with water from your other tanks. This enables them to face any pathogens to which your fish are immune before the added stress

This *Pelvicachromis pulcher* has a parasitic yellow grub (clinostomiasis) growing above its eye. Photo by Ruda Zukal.

of trying to fit into an established social hierarchy.

The ideal situation would be never to move fish from one environment to another, and many commercial breeders maintain strictly "closed" hatcheries, with one-way transport of fish only out of the facilities. Of course, the hobby would be greatly impoverished if everyone followed such a policy, so quarantining affords the next best alternative.

Now even the most fastidious hobbyist occasionally has to deal with disease in his fish. Unavoidable things like power outages, combat injuries, even kismet mean that sometimes fish will get sick anyway, despite your perfect husbandry.

A chilling of the water often brings an attack of Ich (pronounced "ik"), the protozoan white spot disease many people think is named for the icky way the fish look, all pimpled in white, but it is actually short for *Ichthyopthirius*, the oversized name of the dastardly little protozoan responsible for the pimples. The disease can be fatal, but commercial preparations are effective when used according to the instructions. Have some on hand to avoid unnecessary heartache. If you never need it, it was cheap insurance.

Injuries can lead to fungal or bacterial infections. Prompt prophylactic treatment is the best course of action. Remedies range from the time-honored salt bath, to painting the wounds with iodine, to chemical formulations added to the tank water.

Bad luck (or whatever you invoke) is responsible for tumors, organ failures, and other ailments. People die from heart attacks, cancer, and other non-pathogen-caused diseases, and fish have similar problems. Of course, as in humans, proper diet and an unpolluted environment go a long way toward preventing premature death, and if you can bear another repetition, frequent partial water changes are the single most effective removal of toxins, prevention of stress, and deterrent to disease in the aquarium. Coupled with other practices of good aquarium management, they will assure your success and your full enjoyment of your dwarf cichlids for many years to come.

This dwarf cichlid was scraped during shipment and its whole side has erupted with a fungus infection. Most fungus infections are secondary infections caused initially by bacterial infections. Photo by Dieter Untergasser.

Page numbers in **boldface** refer to illustrations.